Clever Kids

Math
Ages 5-7

World Book, Inc.

Chicago London Sydney Toronto

Kids: The answer key is on page 32.

For information on other World Book products,
call **1-800-255-1750, ext. 2238.**

© 1996 World Book, Inc. All rights reserved. This volume may not be reproduced in whole
or in part in any form without prior written permission from the publisher.

World Book, Inc.
525 W. Monroe
Chicago, IL 60661

ISBN: 0-7166-9200-7
LC: 95-61316

Printed in the United States of America

3 4 5 6 7 8 9 10 99 98 97 96

Contents

Puppet Play

YOU NEED:

★ A light- or bright-colored sock that you can draw on (Get permission first!)
★ Markers
★ Two pieces of string
★ Several sheets of plain paper
★ A small, lightweight table

Do you like puppet shows? You and a friend can put on one of your own. All you need are a puppet actor, a stage, and cue cards. Cue cards have all the things the puppet actor will say and do written down on them.

1. First, make your puppet actor. Use the pieces of string to tie off "ears" at the toe of the sock.

2. Draw a mouth, a nose, and eyes with your marker.

3. Slip the sock over your hand. Stuff a little bit of the sock between your thumb and second finger to make a mouth. Meet your puppet actor!

4. Now make some cue cards. With your marker, write one word on each sheet of paper. Write these words:

In	**Out**
Left	**Right**
Back	**Front**
On	**Off**
Bottom	**Top**

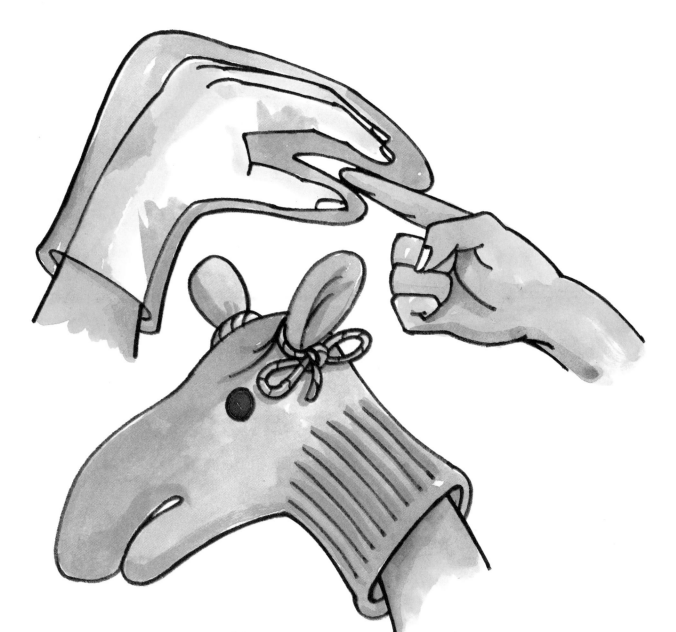

5. For your stage, turn a table onto its side. Sit down behind the table with your puppet actor. Or go behind a chair or sofa.

6. Have your friend hold up the cue cards where you can see them.

7. It's time for your puppet to go on stage! Watch the cue cards, and have your puppet perform the moves. Say: "Puppet moves in," "Puppet goes out," and so on, until you've gone through all the cards.

8. Now let your friend take a turn and help your puppet perform, while you hold the cards.

Counting Around the House

YOU NEED:

★ A large piece of poster board
★ 10 small plastic bags
★ Tape
★ A marker

10 pennies

Have you ever gone on a scavenger hunt?

It's simple and fun! See if you can find these objects

around the house. (Be sure to ask permission first!)

4 paper napkins

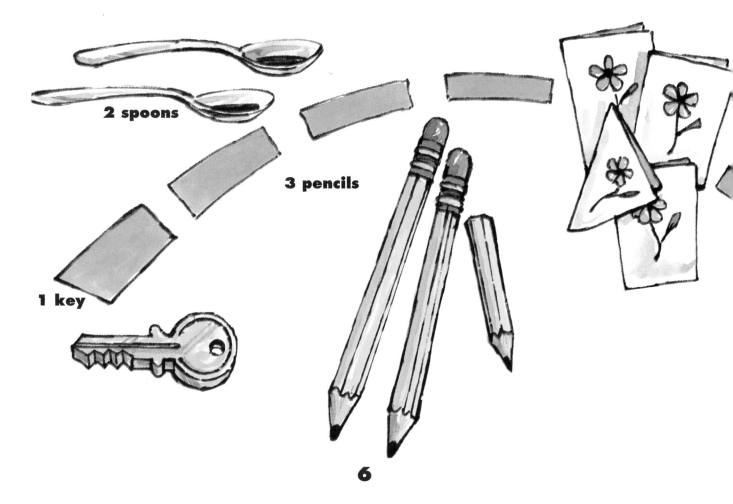

2 spoons

3 pencils

1 key

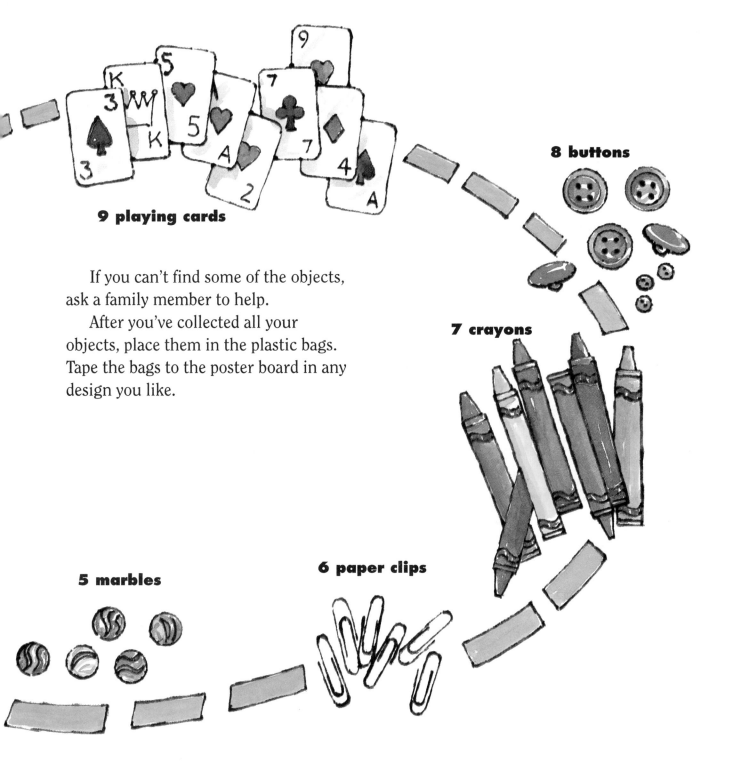

9 playing cards

8 buttons

7 crayons

If you can't find some of the objects, ask a family member to help.

After you've collected all your objects, place them in the plastic bags. Tape the bags to the poster board in any design you like.

5 marbles

6 paper clips

Congratulations! You've "bagged" quite a haul from your hunt. Now it's time to show off all the things you've found around the house.

Which **G**roup?

Take a look in your toy box. There are probably lots of fun things to play with inside. Maybe you see a stuffed animal. Or a ball. Or some toy cars.

Not all your toys look alike. That's easy to see! But maybe they have some things in common. Let's take a closer look.

Are some toys red? Put the red toys in a pile on the floor. Are some round? Make a pile of round toys.

Now put all your toys back in the box.

The pairs of words below are opposites.

soft—hard
small—big
light—dark

Sort your toys into two piles that are different—one for hard toys and one for soft toys. Now put them all back together again.

Make two new piles for small and big. Do the same thing for light and dark. Did you see how the piles changed each time?

Can you think of other pairs of opposite words to describe your toys? Use your new pairs of words to make more piles. Say the word that describes the toys as you build each pile.

Ten-Card Match-Up

YOU NEED:
- ★ 10 playing cards, ace through 10
- ★ A pair of dice

Here's a card game you can play alone or with a friend.

1. Lay 10 playing cards from an ace through a 10 face up in a row. The ace will stand for 1.

2. Toss a die, and count the number of dots on the side that lands up. (Note: A die is one of a pair of dice.) Pick up the card with the same number.

3. Keep tossing the die. If you roll the number of a card you've already picked up, just toss the die again.

4. Soon you will be holding the cards for 1, 2, 3, 4, 5, and 6. Now begin tossing two dice at a time, and add up the numbers on both. Pick up the playing card for that number.

5. Keep tossing the dice until you have picked up all the cards.

You can even play this game with a friend or two.

Lay out 10 cards for each player. Take turns rolling the dice.

Whoever picks up all the cards first is the winner!

Shape Twister

YOU NEED:
- ★ Several sheets of construction paper
- ★ A pencil or marker
- ★ A pair of scissors
- ★ Tape
- ★ Some friends to play with

Here's a game to play with your friends that has a special "twist"!

First, you need circles, squares, and triangles to play.

These shapes and their names are shown below.

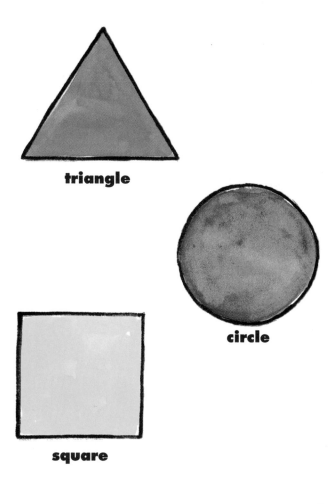

triangle

circle

square

1. Copy the shapes—only larger—on sheets of construction paper. Make three or four copies of each shape.

2. Cut out the shapes; hold them over your head and drop them.

3. Tape the shapes to the floor where they land.

4. Pick one of your friends to be the "caller." The rest of the players should stand around the shapes. Then the caller has to think up directions for each player, like:
   - ★ **"Put your elbow on a square."**
   - ★ **"Place your chin on a circle."**
   - ★ **"Touch your right knee to a triangle."**

"Touch your nose to a triangle."

"Place both hands on a circle."

"Put your left foot on a circle and
your right foot on a square."

Shape **S**earch

YOU NEED:

★ A large paper bag

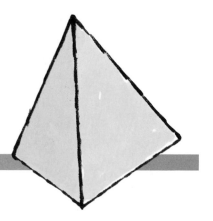

pyramid

Have you ever noticed that some shapes are flat and some are not? Triangles and circles are flat. The shapes you see on this page are solid.

cylinder

Take a look around your house, and see if you can find things that have the same shape as in the pictures.

When you find one of these objects, place it in your paper bag. (Be sure to ask permission before you collect your items.)

Stop searching after you've found 15 things.

Now reach into the bag without looking. Feel one of the items. What shape is it? Try to tell the shape of each thing in the bag by touching it.

Next, put back all the things you collected. As you go from room to room, look for something that has a different shape from any of the shapes in the picture.

sphere

Time's Up!

YOU NEED:
- ★ A clock or watch with a second hand
- ★ Someone to time you
- ★ A banana

"Just a second." People say that when they mean almost no time

at all. Do you know how long a second is?

It's about the time it takes to say "one-Mississippi."

Sixty seconds make up a minute. So can you guess how long a minute takes?

Find out. First, guess how many times you can clap in 1 minute. Then ask someone to time you as you clap. Have the person look at a clock while you clap the number of times that you guessed. Be sure to have your timer tell you when a minute is up.

Did you clap for the whole minute?

Now guess how many times you can clap in 30 seconds. Clap that number of times. But first, remind the timer to tell you when the 30 seconds are up.

Now have the timer watch the clock as you peel a banana. How much time did you use?

How long do you think it takes you to make your bed? Or comb your hair?

Have your timer help you find out the time it takes you to:

- ★ **Brush your teeth**
- ★ **Drink a glass of water**
- ★ **Put on your shoes**
- ★ **Print your name**

13

Pinkies, **P**iggies, and **T**oys

YOU NEED:

★ 6 toy figures

Use your fingers, toes, and toy figures to help you find the endings

to these stories.

Egg Hunt!

"Cluck, cluck," said Red Hen. "Farmer Brown told me I have to lay 12 eggs!"

"Cluck," said Brown Hen. "I'd like to help, but I can only give you 5 eggs. How many do you have so far?"

"I have 4 eggs already," replied Red Hen. "And I think I can lay 1 more, too!"

"You can have 1 of mine, Red Hen," chimed in Speckled Hen.

Would Red Hen have 12 eggs to give to Farmer Brown?

Which Little Piggy?

Count a toe for each piggy that Papa Pig mentions.

"This morning I had 6 clean piggies," Papa Pig snorted. "Since then, 2 have played in the dirt, and 1 rolled in dust. And 2 others sat in a mud puddle."

Did any piggies stay clean?

Birthday Treats

Count a toy for each person at the party. Use your fingers and toes to count the treats.

Bea asked all her friends to a party. Everyone brought treats. Luis brought 4 candied apples. Aiko and Mai each came with 3 chocolate-covered bananas. Maurice showed up with 4 cans of punch, and Consuela arrived with 2 boxes of juice.

Did each person at the party get at least 1 sweet? Did everyone have a drink?

Button, **B**utton, **W**here **A**re the **B**uttons?

YOU NEED:

★ 10 small buttons

Sam keeps losing the buttons on his shirt.

Would you do him a favor and put your real buttons on his shirt?

Put one on each cuff and put one on each pocket. Put the

others in a straight line down the front of his shirt.

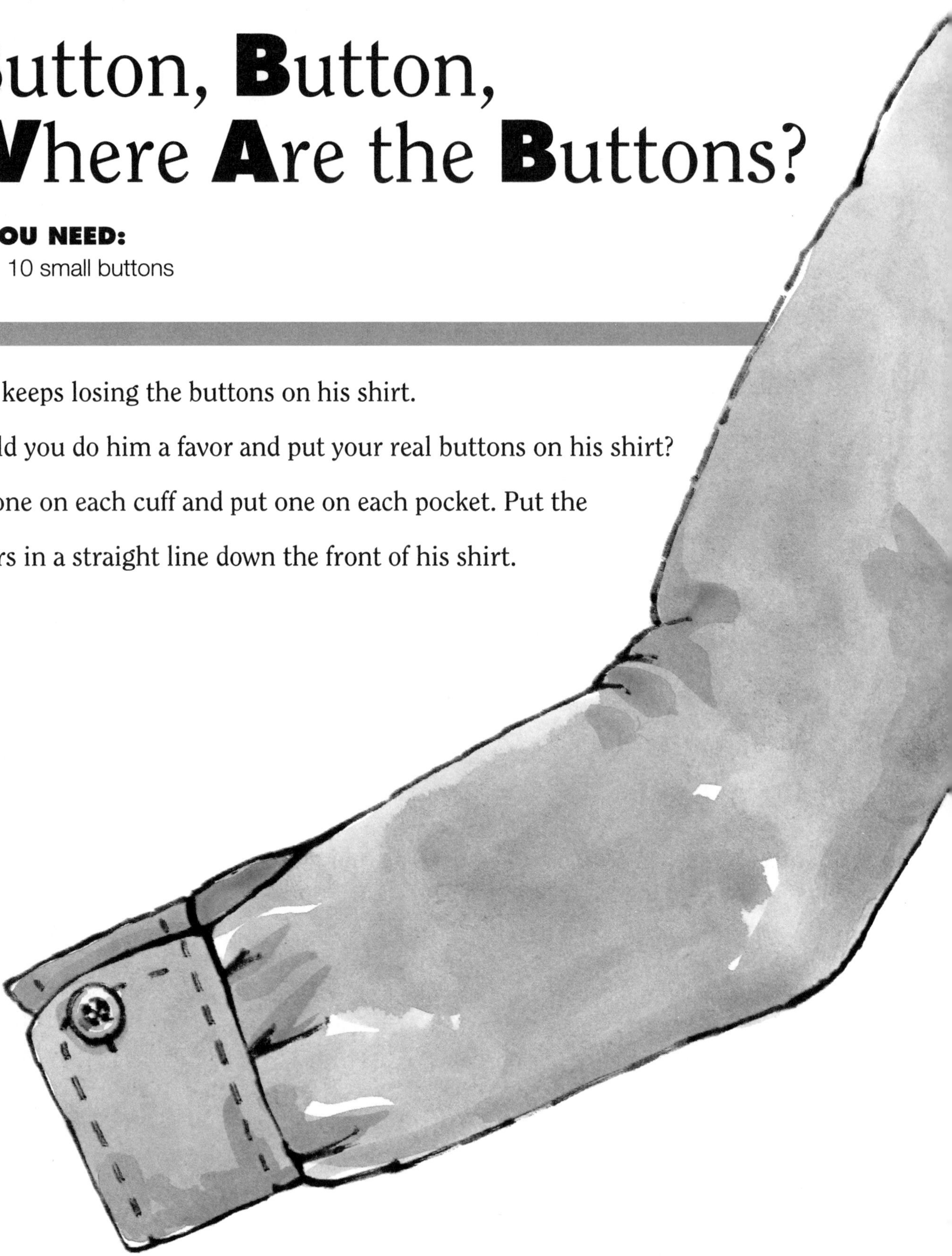

Wait a minute! Sam's puppy chewed on his right-sleeve cuff. Take away the button on that sleeve's cuff. How many buttons are left?

Poor Sam! Now he lost some buttons playing baseball. When he slid into home, the buttons fell off his pockets. Take away the buttons on Sam's pockets. How many buttons are left?

Oh, no! Sam snagged his left sleeve on a tree branch, and another button fell off! Remove the button on Sam's left cuff. Now how many buttons are left?

Sam came home and his mother asked, "Sam, where are your buttons?"

Sam just shrugged. Then he pulled off his shirt, and buttons flew every which way! Take away four buttons from Sam's shirt front. How many buttons does Sam's shirt have now?

Put back all of Sam's buttons. Now make up another story about Sam losing them. Take away the buttons as you tell your story. Then figure out how many buttons are left.

Over the **T**op

YOU NEED:

★ Two 1-cup measuring cups
★ Uncooked macaroni
★ Two paper lunch bags
★ A friend to play with
★ An empty 1/2-gallon milk carton (rinsed out)
★ A 2-quart saucepan
★ A sink

Welcome to the math lab! This is where math scientists test their ideas.

It's also a place where they solve problems and work out math puzzles.

You can be a math scientist, too, with some

simple experiments of your own.

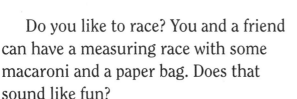

Do you like to race? You and a friend can have a measuring race with some macaroni and a paper bag. Does that sound like fun?

1. Ask your friend to guess how many cupfuls of macaroni will fill a lunch bag.

2. Make your own guess.

3. Make sure that both you and your friend each have your own paper bag, lots of macaroni, and a cup.

4. Race to see how fast you and your friend can fill your paper bag with the macaroni. Be sure and count the cupfuls as you fill the bags.

Who won the race? Whose guess was closest to the right answer?

Now let's try an experiment with a milk carton and a saucepan.

Start by taking a close look at the milk carton and saucepan. Will the carton hold more water than the pan? Will the pan hold more? Or will the milk carton and saucepan hold the same amount?

Take your best guess. Then think about why you made your choice. Now let's see if you were right!

1. Fill the milk carton with water.

2. Place the saucepan in the sink.

3. Pour the water from the carton into the saucepan.

Did the water go over the top of the pan? It shouldn't have, because the carton and pan hold the same amount of water. But the carton is taller, and the pan is wider. Did that fool you at first?

While we're here at the math lab, let's do another test.

1. Guess how many cupfuls of water will fill the saucepan.

2. Measure the water into the saucepan until you reach the number you guessed.

Did you fill the carton? Or did the water go over the top? How many cupfuls of water filled the saucepan?

Jewelry to Wear . . . and Eat!

YOU NEED:

★ Licorice whips
★ Cereal O's
★ O-shaped hard candies, such as Life Savers®
★ Tube-shaped licorice candies, such as Snaps® or Licorice Bites®

If you can string beads, then you can make jewelry that tastes as good as it looks. Sounds like some pretty "tasty" fun, doesn't it? Just don't forget that you have to finish making your jewelry before you eat it!

Look at the picture on this page and the next. It shows cereal and candy lined up in a certain order, or pattern. First there is one licorice candy. The licorice candy is followed by:

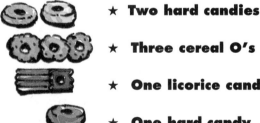

★ **Two hard candies**

★ **Three cereal O's**

★ **One licorice candy**

★ **One hard candy**

★ **Two cereal O's**

Then the pattern starts all over again, with one licorice candy.

Follow this pattern when you make your own necklace. Now let's get started.

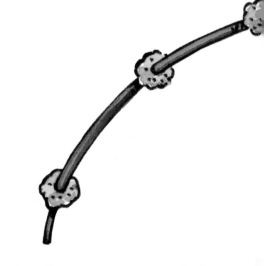

1. Lay out all your candies and cereal in front of you.

2. Pick up a licorice whip with one hand and string the candy and cereal onto your whip with the other. Be sure to follow the pattern!

3. As you work, repeat the pattern aloud. Say: "One licorice candy, two hard candies, three cereal O's," and so on.

4. Repeat the pattern again and again until you cover the whole whip except for a finger's length at each end.

5. Tie the ends together. Now you're ready to wear your new necklace!

Would you like to make a bracelet to go with your necklace? Follow the same steps you did for the necklace. This time, though, use only half a length of whip.

Make up your own pattern. Be sure to follow your pattern exactly.

When you're finished, you can wear your new necklace and bracelet. Or you can give them away to a friend or family member. Or you can eat them. Yum!

The Penny Store

YOU NEED:

★ 20 pennies
★ 6 cardboard strips
★ 4 nickels
★ 2 dimes

Place one of your real pennies over each picture penny

at the top of the next page. Now the Penny Store is open for

business, and you're ready to shop!

Count the penny pictures below something you want. Take that number of pennies from the top of the page. Place them over the picture pennies below what you want.

How many pennies are left? What can you buy with them? Find out by counting the picture pennies below other things at the store.

You can spend nickels and dimes at the store, too.

1. Cut cardboard strips as long as the rows of picture pennies at the tops of the pages.

2. Tape a dime on each strip. Make two more strips. Cut them in half.

3. Tape a nickel on each of these half-strips.

Ten pennies equal a dime. So cover groups of 10 pennies with the long strips. Five pennies equal a nickel. So cover groups of 5 pennies with the half-strips.

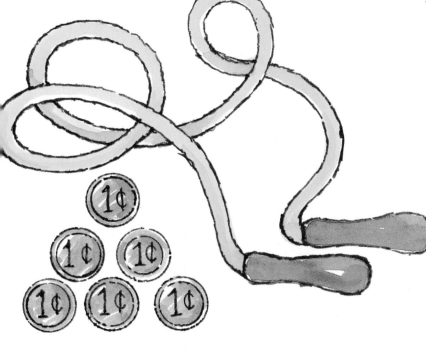

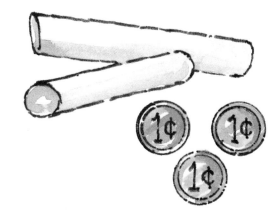

Bank Roll

YOU NEED:

★ Two rolls of pennies, for a total of 100 pennies
★ An empty shoebox

Here's a fun way to start a bank account. Pretend the empty

shoebox is the bank. Now it's up to you to make sure you

"deposit" 100 pennies in your bank.

1. Place the shoebox on the floor and turn it on its side.

2. Lie on the floor a few feet away from the shoebox. Take your two rolls of pennies with you.

3. Roll the coins one by one into the shoebox. (This works better on a bare floor than it does on a rug.) Count them as you roll them. Stop when you reach 100.

Now stack your coins. How many coins can you put in one stack before it topples over? Build stacks of 10 coins each. How many of these stacks can you make with your 100 coins?

Next, build stacks of 5 coins each. How many of these stacks can you make from 100 coins?

I Spy Two

YOU NEED:

★ A shopping bag
★ Permission to collect pairs of objects around the house
★ Someone to watch the time

Pairs are things that come in twos.

You can find pairs hiding just about everywhere—

in cabinets, in drawers, and in closets.

Can you spy a pair of salt and pepper shakers in the kitchen cabinet? Uncover a pair of socks in a chest of drawers? Detect a pair of shoes in the closet?

Look around your house for all these kinds of pairs and more. Bring your shopping bag with you. See how many pairs you can find in five minutes. Ask someone to tell you when the time is up.

1. Place each pair you find in your shopping bag. As you do, count each item.

2. Say the first number in the pair softly and the second number loudly, like this: "1, **2.** 3, **4.** 5, **6.** . . ." and so on.

3. After your five minutes are up, return all the pairs to where they belong. Again, count each item. But this time, say only every other number aloud, like this: "2, 4, 6. . . ."

4. Keep count by twos until your bag is empty.

Heart and Flower Cutouts

YOU NEED:
- ★ Tracing paper
- ★ A pencil
- ★ Scissors
- ★ Sheets of construction paper
- ★ Tape

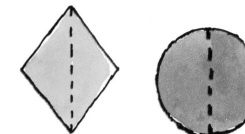

Would you like to be a math magician? It's easy and fun!

All you need to do is learn some tricks with symmetry. This is how

you say the word: SIM uh tree. What does this new word mean?

The square, diamond, and circle at the top of this page will show you. Each of these shapes has a dotted line down its center. See how the parts on either side of the lines are the same. Why? Because the shapes have symmetry. Each side is the same as the other.

Now use the following trick to make other shapes with symmetry.

1. On a sheet of tracing paper, trace the drawings of the daisy and clover above and cut them out.

2. Fold a sheet of construction paper in half.

3. Place the straight edges of your cutouts along the fold line of the construction paper and trace them.

4. Cut around the outlines you just traced.

5. Open up your new cutouts to discover their shape.

Now that you're a math magician, try another trick that's even more amazing.

1. Trace this shape onto tracing paper and cut it out.

2. Fold a sheet of construction paper in half lengthwise. Cut along the fold line. You now have two long sheets of paper.

3. Fold one of the long sheets of paper in half lengthwise. Then fold the sheet in half lengthwise again. Then fold it in half lengthwise once more.

4. Place the straight edge of your cutout along the fold line of your long, thin paper. Be sure that part of your cutout touches the other side, too.

5. Trace the cutout and cut around the outline. Leave some of the fold on both sides of the outline uncut.

6. Unfold your new cutout. Like magic, you have a chain of hearts!

If you'd like, you can make another chain of hearts with the second sheet of construction paper. Tape your two chains together to make a longer chain.

Star Gazing

YOU NEED:
★ Tracing paper
★ A pencil
★ Crayons

Look up in the sky at night. There must be millions of stars twinkling everywhere! They look like tiny white dots on a big dark blanket.

Now look at the picture on the next page. It has 100 stars on it. Let's see how you can connect the stars to form a constellation (KAHN stuh LAY shuhn). A constellation is a group of stars. In ancient times, people often named these groups after animals.

1. Put a sheet of tracing paper over the star picture. With your pencil, connect the dots from 1 to 100 on the tracing paper. Count aloud as you go. Say: "1, 2, 3, 4," and so on.

2. When you've connected all the dots, what kind of creature does your constellation look like?

3. Use your crayons to color your constellation.

2
3
4
5
6
1
7
8
100 99 98 97 9
96
95 10
94 11
93
92 12
91 13
90 14
89 15
88
85 86 87 16
83 84 17
82 18
81 19
80
79
78 20
77
76
75 21
74
73 57 56 39
58 55 48 40 38 30 22
54 49 41 37 31 29 23
2 50 47 42 28
59 43 36 32 24
71 53 27
60 51 52 44 35 34 33 26 25
61 46 45
62
70 63
69 64
68 67
66 65

Bingo 1000

YOU NEED:

- ★ Sheets of plain paper
- ★ Markers
- ★ A large bowl
- ★ Scissors
- ★ Construction paper
- ★ Friends to play with
- ★ Pennies or old buttons

Do you like to play bingo? You and your friends

can make your own bingo game cards and have fun

playing this number game.

1. Fold sheets of plain paper in half four times.

2. Open up the sheet and draw lines where the folds are. You now have 16 squares on your paper.

3. Look at the sample bingo cards on page 30. Copy their numbers onto your sheets. Say the numbers aloud as you write them.

4. Make two copies of each sheet. Cut apart the numbers on the second copy. Put all the numbers you cut out into a bowl.

5. Find something around your house to use as your bingo markers, like pennies or old buttons. Or make your own markers by cutting small circles out of construction paper.

Now you and your friends are ready to play!

Ask one of the players to pick numbers from the bowl and call them out. Do you see any of these numbers on your sheet? Place markers over them. The first player to cover a row of numbers up or down shouts, "Bingo!"

Answer **K**ey

Pinkies, Piggies, and Toys
Pages 14-15

★ No, Red Hen would have only 11 eggs.

★ Yes, one piggy stayed clean.

★ Each person got at least one sweet.

Everybody had a drink.

Bank Roll
Page 24

You can make 20 stacks of 5 coins each.

Star Gazing
Pages 28-29

The picture is a dinosaur.